DINOSAUR DIG

Dinosaur Combat

Unearth the secrets behind dinosaur fossils

QEB Publishing

Rupert Matthews

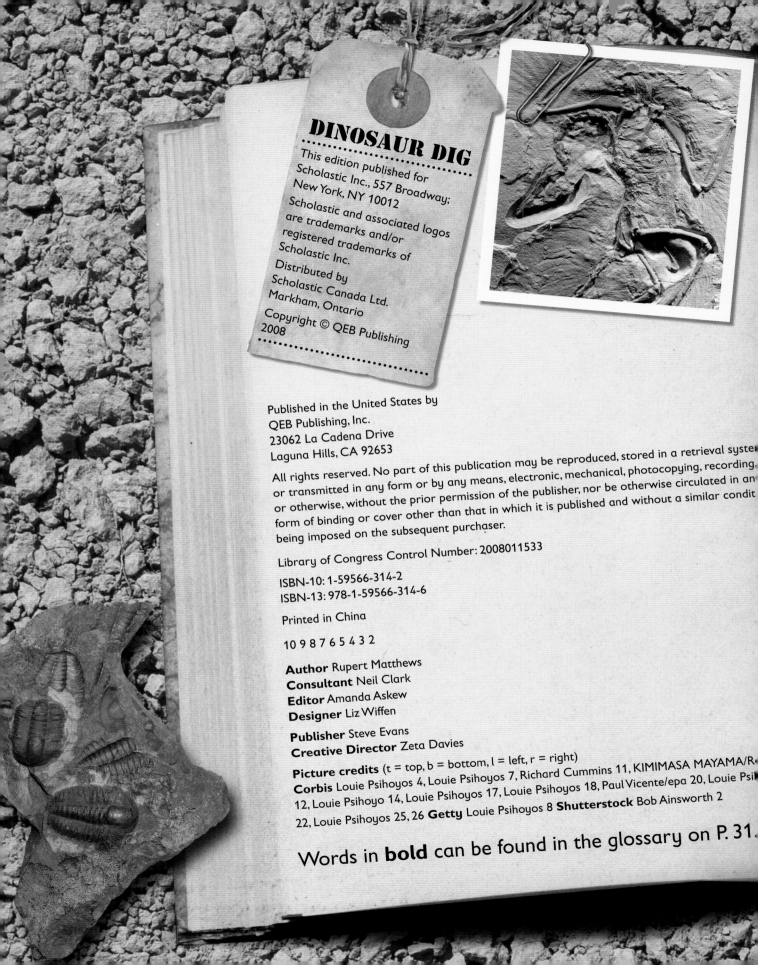

DINOSAUR DIG

This edition published for
Scholastic Inc., 557 Broadway;
New York, NY 10012

Scholastic and associated logos
are trademarks and/or
registered trademarks of
Scholastic Inc.

Distributed by
Scholastic Canada Ltd.
Markham, Ontario

Copyright © QEB Publishing
2008

Published in the United States by
QEB Publishing, Inc.
23062 La Cadena Drive
Laguna Hills, CA 92653

Library of Congress Control Number: 2008011533

ISBN-10: 1-59566-314-2
ISBN-13: 978-1-59566-314-6

Printed in China

10 9 8 7 6 5 4 3 2

Author Rupert Matthews
Consultant Neil Clark
Editor Amanda Askew
Designer Liz Wiffen

Publisher Steve Evans
Creative Director Zeta Davies

Picture credits (t = top, b = bottom, l = left, r = right)
Corbis Louie Psihoyos 4, Louie Psihoyos 7, Richard Cummins 11, KIMIMASA MAYAMA/R 12, Louie Psihoyo 14, Louie Psihoyos 17, Louie Psihoyos 18, Paul Vicente/epa 20, Louie Psi 22, Louie Psihoyos 25, 26 **Getty** Louie Psihoyos 8 **Shutterstock** Bob Ainsworth 2

Words in **bold** can be found in the glossary on P. 31.

CONTENTS

DINO GUIDE
Learn how to pronounce the dinosaurs' names, find out their length and weight, and discover what they ate.

DINOSAUR DIG

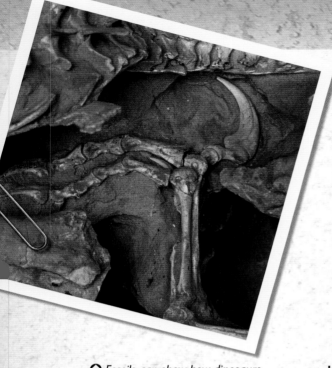

Dinosaurs **were a group of** reptiles **that lived millions of years ago. They became** extinct, **or died out, about 65 million years ago.**

Some dinosaurs were plant eaters, and others were hunters, or meat eaters. Scientists called **paleontologists** (pay-lee-on-toll-oh-jists) look at dinosaur remains, called **fossils**, to learn about dinosaurs.

⬤ *Fossils can show how dinosaurs fought. This fossil shows a Velociraptor (vel-oss-ih-rap-ter) claw stuck in the body of Protoceratops (pro-toe-ser-ah-tops).*

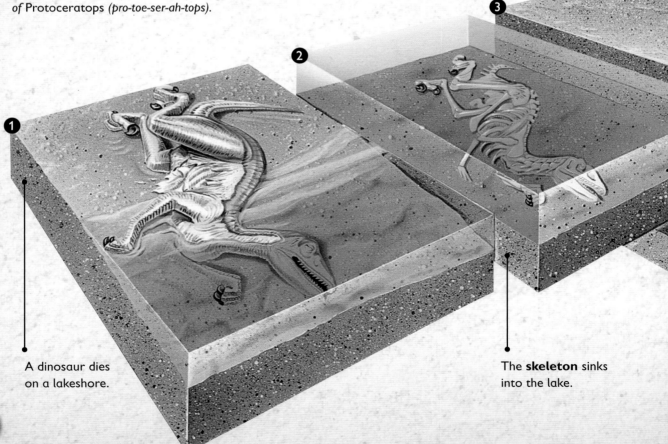

1 A dinosaur dies on a lakeshore.

2

3

The **skeleton** sinks into the lake.

Fossils help scientists to understand what dinosaurs looked like and how they behaved. They can even tell which dinosaurs were involved in fights, or combat, by studying their horns, teeth, or damaged bones.

How big were dinosaurs?

Every dinosaur is compared to an average adult, about 5 ft. 2 in. (48 m) in height, to show just how big they really were.

◐ *When a plant or animal dies, it usually rots away completely. However, in special conditions, parts of it can become fossilized.*

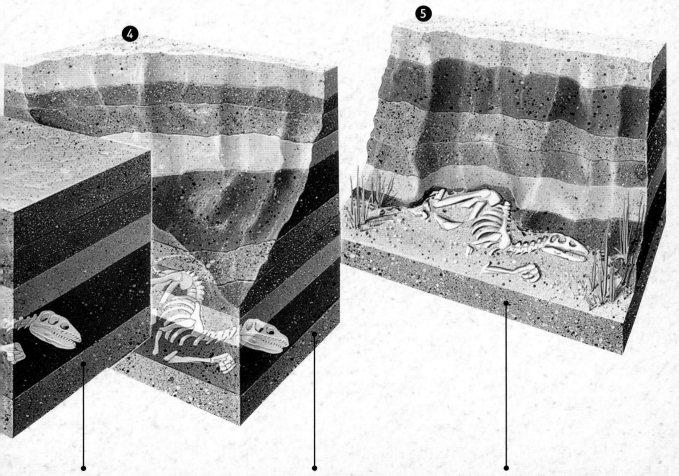

4

5

Layers of mud settle over the skeleton. The mud and bones gradually turn into stone.

The rock wears away, or **erodes**.

As more rock erodes, the skeleton is revealed.

THE HUNTERS

Hunting dinosaurs had many different weapons for attacking their prey, such as teeth and claws.

Some hunters worked alone, but others lived in groups called **packs**. If a pack of small hunters attacked a larger hunter, the fight would be dramatic.

Herrerasaurus (he-ray-ra-saw-rus) was one of the largest hunters of the late Triassic Period. With sharp teeth, it was very ferocious. *Herrerasaurus* could catch smaller dinosaurs, such as *Eoraptor* (ee-oh-rap-ter), with a single bite. If *Eoraptor* formed a pack, they might stand a chance of survival.

DINOSAUR DIG

Eoraptor

....

Herrerasaurus

....

WHERE: Argentina, South America

....

PERIOD: 225 million years ago in the late Triassic

....

DIG SITE

◗ *A pack of* Eoraptor *attack a much larger* Herrerasaurus. *The larger dinosaur is more powerful, but a pack of smaller dinosaurs would be able to fight together.*

Eoraptor

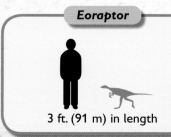

3 ft. (91 m) in length

A fossilized *skull* of Herrerasaurus shows the curved-back teeth that helped the dinosaur to grip struggling prey.

Herrerasaurus

10 ft. (3 m) in length

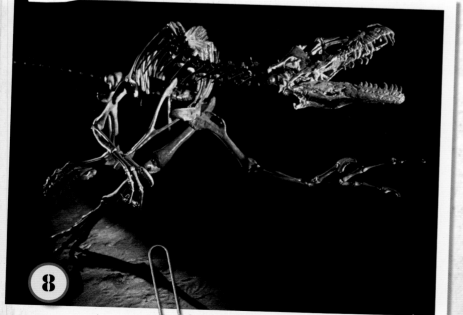

CARRION EATERS

Hunting dinosaurs did not always have to find and kill their prey. Sometimes they found a meal just waiting to be eaten.

A dead body, or **carcass**, that has begun to rot is called **carrion**. Some meat eaters had an extremely good sense of smell and sight to help them find carrion. They rarely hunted at all. However, even the strongest hunters would feed on carrion if they came across it.

DINOSAUR DIG
Dromaeosaurus

WHERE: Canada, North America

PERIOD: 75 million years ago in the late Cretaceous

DIG SITE

◑ Dromaeosaurus (drom-ee-oh-saw-rus) was equipped with excellent killing weapons—fanglike teeth and very sharp claws.

◔ A group of hunters squabbles over the carcass of a rhynchosaur. Although Dromaeosaurus could easily catch prey, it would also feed on carrion.

WOW!

Carrion eaters need an excellent sense of smell so that they can find carcasses.

Dromaeosaurus

6.5 ft. (2 m) in length

TAIL SPIKES

Plant-eating dinosaurs needed to be able to protect themselves from hunters in order to survive.

One group of dinosaurs, called **stegosaurs**, developed long, sharp spikes on their tail. *Kentrosaurus* (ken-troe-saw-rus) was a stegosaur with upright plates of bone along its back, as well as sharp spikes along its tail.

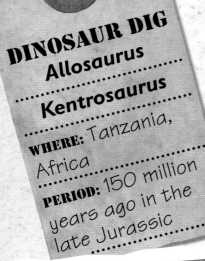

DINOSAUR DIG

Allosaurus ....

Kentrosaurus ....

WHERE: Tanzania, Africa

PERIOD: 150 million years ago in the late Jurassic ....

DIG SITE

If *Kentrosaurus* faced a large hunter, such as *Allosaurus* (al-oh-saw-rus), it would easily be defeated. The only chance *Kentrosaurus* had of surviving was to hit *Allosaurus* with its tail spikes. This would injure the hunter, and *Kentrosaurus* could escape.

WOW!

The bone spikes of *Kentrosaurus* would have been covered in shiny horn with extremely sharp points, making them excellent weapons.

◗ Allosaurus *attacks* Kentrosaurus. *The hunter is more than twice as large, so* Kentrosaurus *has to use all its power to escape from* Allosaurus.

◐ *A fossil skeleton of Allosaurus shows how it might stride forward while hunting. The head could lunge out to bite prey.*

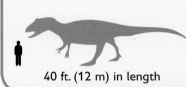

Allosaurus

40 ft. (12 m) in length

Kentrosaurus

16.5 ft. (5 m) in length

THE CHASE

Some smaller plant eaters relied on speed to escape from danger.

They had no weapons and were fairly weak, so they would run away as soon as they saw a hunter. However, many meat eaters were also able to run very quickly, so they would chase the smaller dinosaur.

If a plant eater, such as *Hypsilophodon* (hip-see-loff-oh-don), could run faster than a hunter, such as *Deinonychus* (die-no-ni-kiss), it would escape. If it could not, then it would fall victim and end up as a meal for the hunter.

DINOSAUR DIG
Deinonychus
....
Hypsilophodon
....
WHERE: Montana, North America
....
PERIOD: 100 million years ago in the early Cretaceous
....

DIG SITE

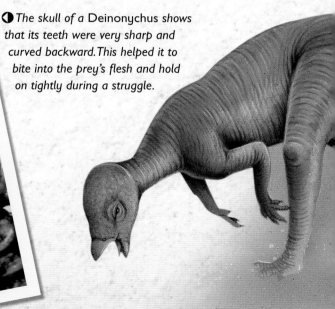

◑ *The skull of a Deinonychus shows that its teeth were very sharp and curved backward. This helped it to bite into the prey's flesh and hold on tightly during a struggle.*

12

● *If Deinonychus attacked a herd of* Hypsilophodon, *they would panic and spread out. One of them would be slower than the rest and would easily be injured by the sharp claws of Deinonychus.*

WOW!

Some scientists believe that Deinonychus may have been covered in feathers, but others think that its skin was scaly (see PP. 18–19).

Hypsilophodon

8 ft. (2.4 m) in length

Deinonychus

10 ft. (3 m) in length

HUNTING ALONE

A hunter working alone would have avoided attacking a large plant eater.

It would have been difficult for meat eaters, such as *Ceratosaurus* (se-rat-oh-saw-rus), to attack *Brachiosaurus* (brack-ee-oh-saw-rus) because it was so large. Although *Brachiosaurus* had no weapons, such as sharp teeth or claws, it could stamp or kick with great force. *Ceratosaurus* would need to take *Brachiosaurus* by surprise, or be very lucky, to win the combat.

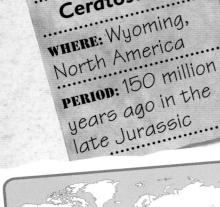

DINOSAUR DIG
Brachiosaurus
Ceratosaurus
WHERE: Wyoming, North America

PERIOD: 150 million years ago in the late Jurassic

DIG SITE

◗ *This famous skeleton of a* Brachiosaurus *from a museum in Berlin, Germany, is the largest mounted dinosaur skeleton in the world. The fossilized skeleton had several bones missing, which were replaced with fossil bones taken from other similar dinosaurs.*

Brachiosaurus

80 ft. (24.4 m) in length

WOW!

Brachiosaurus had nostrils on top of its head. This enabled the dinosaur to make a noise that may have been used to communicate with other dinosaurs.

⬤ *Ceratosaurus prepares to attack an **adult** Brachiosaurus. Hunters would probably have preferred to avoid such large individuals and would attack younger animals instead.*

Ceratosaurus

20 ft. (6.1 m) in length

EASY PREY

Most hunters preferred to find an easier meal than fighting a fully grown sauropod. Young sauropods were easier to kill.

Young dinosaurs were smaller and weaker than adults, and they had less experience fighting or escaping from danger.

Megalosaurus (meg-ah-low-saw-rus) was armed with sharp teeth in strong jaws, and had powerful claws on its feet. If it could catch a dinosaur smaller than itself, it would have an easy meal. Old or sick animals were also easier to overcome than healthy adults.

DINOSAUR DIG
Megalosaurus

WHERE: France, Europe

PERIOD: 165 million years ago in the mid-Jurassic

DIG SITE

WOW!

Fossilized sauropod bones have been found covered in scratch and bite marks—probably from the teeth of hunting dinosaurs!

◐ Megalosaurus *prepares to eat a young sauropod that it has killed. It was the most powerful hunter in Europe during the late Jurassic Period.*

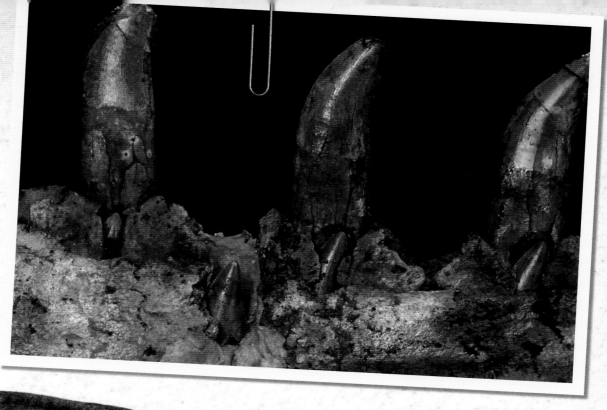

● *The teeth of Megalosaurus were very sharp and curved backward. This would give the dinosaur a firm grip on struggling prey.*

Megalosaurus

30 ft. (9.1 m) in length

FATAL WOUNDS

Deinonychus (die-no-ni-kiss) belonged to a group of ferocious hunters known as raptors.

These fast-moving hunters had a large, curved claw on each of their back legs. This weapon was held off the ground so that it stayed sharp and ready for action.

DINOSAUR DIG

Deinonychus ....

Tenontosaurus ....

WHERE: Oklahoma, North America ....

WHEN: 100 million years ago in the early Cretaceous ....

DIG SITE

◑ *A pack of* Deinonychus *attack* Tenontosaurus. *Scientists have found a fossil showing that* Tenontosaurus *had once been killed by a group of these hunters.*

◑ *This skeleton shows* Deinonychus *leaping forward as though it were about to attack a victim.*

18

A group of *Deinonychus* may have pounced on a larger dinosaur and used their back claws to cause deep wounds. Then they would run off before *Tenontosaurus* (ten-on-toe-saw-rus) could fight back. They would probably wait for their prey to bleed to death, and then move in to feast on the body.

Tenontosaurus

23 ft. (7 m) in length

Deinonychus

10 ft. (3 m) in length

THE AMBUSH

Tarbosaurus (tar-bow-saw-rus) was a large, powerful hunter. It had long, sharp teeth set in jaws that were powered by very strong muscles.

However, it was unable to run very quickly. The best chance it had of killing prey was to ambush it. *Tarbosaurus* would wait in bushes or behind trees, and then leap out on a victim.

Scientists know a lot about *Tarbosaurus* because they have found many fossilized skeletons. Few other dinosaurs from Asia have been found in such numbers, so there must have been many of them around in the late Cretaceous Period.

DINOSAUR DIG
Tarbosaurus

WHERE: Mongolia, Asia

WHEN: 80 million years ago in the late Cretaceous

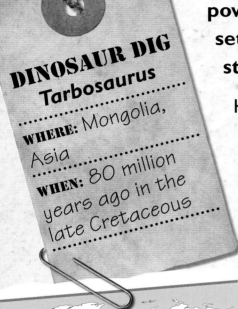

DIG SITE

WOW!

The tiny arms of *Tarbosaurus* were too small to reach its mouth, so scientists are not sure what they were for.

The teeth of Tarbosaurus were smaller than those of its close relative Tyrannosaurus (tie-rann-oh-saw-rus).

◓ The mouth of Tarbosaurus could be opened very wide to reveal its fangs. The wide **gape** and strong jaws show that it may have killed prey by running at them with its mouth open.

Tarbosaurus

40 ft. (12.2 m) in length

21

WEAPONS

When a hunter attacked prey, it would try to avoid any weapons that the plant eater had. The plant eater would do its best to use those weapons to defend itself.

Triceratops (try-ser-ah-tops) had three long, sharp horns on its head to defend itself against hunters. When *Tyrannosaurus* (tie-rann-oh-saw-rus) attacked, *Triceratops* would stab the hunter. If *Tyrannosaurus* became injured, then *Triceratops* would be able to escape.

DINOSAUR DIG

Triceratops

Tyrannosaurus

WHERE: Alberta, North America

WHEN: 70 million years ago in the late Cretaceous

DIG SITE

◑ *A skeleton of Tyrannosaurus shows how it would lunge forward to attack its prey.*

Tyrannosaurus

40 ft. (12.2 m) in length

22

If *Tyrannosaurus* could only make one good bite, it may have stood back to wait for the plant eater to become weak through loss of blood. Then, it would move in to make the kill.

⬤ Tyrannosaurus *battles with* Triceratops *by trying to bite into the soft sides of the plant eater while avoiding its sharp horns.*

Triceratops

30 ft. (9.1 m) in length

THE TAIL CLUB

The armored dinosaurs, or ankylosaurs, **had a unique way of defending themselves from attack.**

The back, sides, head, and tail of *Pinacosaurus* (pin-ah-coe-saw-rus) had a thick armor of bone covered in horns. *Pinacosaurus* also had a heavy bone tail club, which it would use to stop an attacker, such as *Tarbosaurus* (tar-bow-saw-rus), from flipping it over. A blow from the club could seriously injure a hunter.

DINOSAUR DIG

Pinacosaurus ....

Tarbosaurus ....

WHERE: Mongolia, Asia ....

WHEN: 80 million years ago in the late Cretaceous ....

DIG SITE

Pinacosaurus

18 ft. (5.5 m) in length

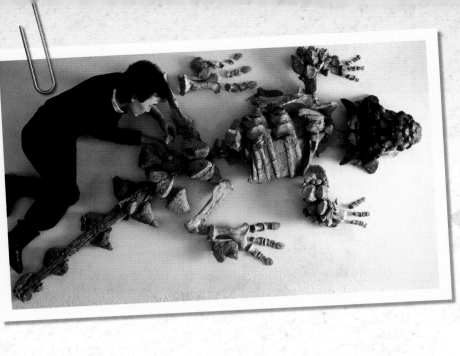

◑ *The fossilized skeleton of an ankylosaur. Many skeletons are found with the bones scattered, so they need to be put back in position to show what the dinosaur looked like.*

WOW!

Ankylosaurs have been found all over the world, except in Africa.

◖ Tarbosaurus *is knocked over by a hit from the tail of Pinacosaurus.* To make a successful attack, Tarbosaurus *had to turn the armored dinosaur over and attack its soft belly.*

Tarbosaurus

40 ft. (12.2 m) in length

If a hunting dinosaur was extremely hungry, it might risk an attack on a plant eater that was ready to defend itself.

Tyrannosaurus (tie-rann-oh-saw-rus) was a powerful killer and may sometimes have become desperate enough to attack an equally strong victim.

Triceratops (try-ser-ah-tops) had a huge, sharp horn growing from its nose, which could cause a serious wound to *Tyrannosaurus*.

DINOSAUR DIG

Triceratops

Tyrannosaurus

WHERE: Montana, North America

WHEN: 70 million years ago in the late Cretaceous

DIG SITE

Tyrannosaurus

40 ft. (12.2 m) in length

◖ *Before* Triceratops *could use its horns in defense,* Tyrannosaurus *attacks.* Tyrannosaurus *uses its claws to wound and kill* Triceratops.

◑ *The teeth of Tyrannosaurus were only lightly fixed to the jaw and often broke off, so Tyrannosaurus constantly grew new teeth.*

WOW!

Triceratops means "three horns on face." The horns on the top of its head were more than 3 ft. (1 m) in length.

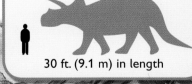

Triceratops

30 ft. (9.1 m) in length

DINO GUIDE

Coelophysis
PRONUNCIATION
see-low-fye-sis
LENGTH 10 ft. (3 m)
WEIGHT 75–80 lbs. (34–36 kg)
DIET Small animals

Efraasia
PRONUNCIATION
ef-rah-see-ah
LENGTH 23 ft. (7 m)
WEIGHT 1,300 lbs. (590 kg)
DIET Plants

Eoraptor
PRONUNCIATION
ee-oh-rap-tor
LENGTH 3 ft. (91 cm)
WEIGHT 7–30 lbs. (3–14 kg)
DIET Small animals

Herrerasaurus
PRONUNCIATION
he-ray-ra-saw-rus
LENGTH 10 ft. (3 m)
WEIGHT 450 lbs. (204 kg)
DIET Animals

Melanorosaurus
PRONUNCIATION
mel-an-or-oh-saw-rus
LENGTH 33 ft. (10 m)
WEIGHT 1 ton (907 kg)
DIET Plants

Pisanosaurus
PRONUNCIATION
peez-an-oh-saw-rus
LENGTH 3 ft. (91 cm)
WEIGHT 6.5 lbs. (3 kg)
DIET Plants

Plateosaurus
PRONUNCIATION
plat-ee-oh-saw-rus
LENGTH 26 ft. (8 m)
WEIGHT 1 ton (907 kg)
DIET Plants

Procompsognathus
PRONUNCIATION
pro-comp-sog-nay-thus
LENGTH 4 ft. (1.2 m)
WEIGHT 4.5–7 lbs. (2–3.2 kg)
DIET Small animals

Riojasaurus
PRONUNCIATION
ree-oh-ha-saw-rus
LENGTH 33 ft. (10 m)
WEIGHT 1 ton (907 kg)
DIET Plants

Staurikosaurus
PRONUNCIATION
store-ick-oh-saw-rus
LENGTH 6.5 ft. (2 m)
WEIGHT 65 lbs. (29.5 kg)
DIET Small animals

JURASSIC PERIOD
206 TO 145 MILLION YEARS AGO

Allosaurus
PRONUNCIATION
al-oh-saw-rus
LENGTH 40 ft. (12.2 m)
WEIGHT 1.5–2 tons (1.4–1.8 tonnes)
DIET Animals

Haplocanthosaurus
PRONUNCIATION
hap-low-kan-thoe-saw-rus
LENGTH 72 ft. (22 m)
WEIGHT 20 tons (18 tonnes)
DIET Plants

Brachiosaurus
PRONUNCIATION
brack-ee-oh-saw-rus
LENGTH 80 ft. (24.4 m)
WEIGHT 50 tons (45 tonnes)
DIET Plants

Kentrosaurus
PRONUNCIATION
ken-troe-saw-rus
LENGTH 16.5 ft. (5 m)
WEIGHT 2 tons (1.8 tonnes)
DIET Plants

Camptosaurus
PRONUNCIATION
kamp-toe-saw-rus
LENGTH 20 ft. (6.1 m)
WEIGHT 1–2 tons (0.9–1.8 tonnes)
DIET Plants

Lesothosaurus
PRONUNCIATION
le-so-toe-saw-rus
LENGTH 3 ft. (91 cm)
WEIGHT 4–7 lbs. (1.8–3.2 kg)
DIET Plants

Ceratosaurus
PRONUNCIATION
se-rat-oh-saw-rus
LENGTH 20 ft. (6.1 m)
WEIGHT 1,500–2,000 lbs. (680–907 kg)
DIET Animals

Megalosaurus
PRONUNCIATION
meg-ah-low-saw-rus
LENGTH 30 ft. (9 m)
WEIGHT 1 ton (907 kg)
DIET Plants

Euhelopus
PRONUNCIATION
you-hel-oh-puss
LENGTH 30–50 ft. (9–15 m)
WEIGHT 10–25 tons (9–23 tonnes)
DIET Plants

Supersaurus
PRONUNCIATION
soo-per-saw-rus
LENGTH 100–130 ft. (30.5–39.6 m)
WEIGHT 30–50 tons (27–45 tonnes)
DIET Plants

CRETACEOUS PERIOD
145 TO 65 MILLION YEARS AGO

Deinonychus

PRONUNCIATION
die-non-ee-kuss
LENGTH 10 ft. (3 m)
WEIGHT 130 lbs. (60 kg)
DIET Small animals

Dromaeosaurus

PRONUNCIATION
drom-ee-oh-saw-rus
LENGTH 6.5 ft. (2 m)
WEIGHT 55 lbs. (25 kg)
DIET Animals

Hypsilophodon

PRONUNCIATION
hip-see-loff-oh-don
LENGTH 8 ft. (2.4 m)
WEIGHT 45–90 lbs. (20–41 kg)
DIET Plants

Pinacosaurus

PRONUNCIATION
pin-ah-coe-saw-rus
LENGTH 18 ft. (5.5 m)
WEIGHT 1–2 tons (0.9–1.8 tonnes)
DIET Plants

Protoceratops

PRONUNCIATION
pro-toe-ser-ah-tops
LENGTH 6.5 ft. (2 m)
WEIGHT 330–550 lbs. (150–250 kg)
DIET Plants

Tarbosaurus

PRONUNCIATION
tar-bow-saw-rus
LENGTH 40 ft. (12.2 m)
WEIGHT 4 tons (3.6 tonnes)
DIET Large animals

Tenontosaurus

PRONUNCIATION
ten-on-toe-saw-rus
LENGTH 23 ft. (7 m)
WEIGHT 1 ton (907 kg)
DIET Plants

Triceratops

PRONUNCIATION
try-ser-ah-tops
LENGTH 30 ft. (9.1 m)
WEIGHT 5–8 tons (4.5–7.3 tonnes)
DIET Plants

Tyrannosaurus

PRONUNCIATION
tie-rann-oh-saw-rus
LENGTH 40 ft. (12.2 m)
WEIGHT 6 tons (5.4 tonnes)
DIET Large animals

Velociraptor

PRONUNCIATION
vel-oss-ee-rap-ter
LENGTH 6.5 ft. (2 m)
WEIGHT 45–65 lbs. (20–29.5 kg)
DIET Small animals

GLOSSARY

Adult An animal that is fully grown.

Ankylosaur A type of dinosaur that had armor across its back and other parts of its body.

Carcass The body of a dead animal.

Carrion Meat from a dead animal that the hunter has not killed itself.

Cretaceous The third period of time in the age of the dinosaurs. The Cretaceous began about 145 million years ago and ended about 65 million years ago.

Dinosaur A type of reptile that lived millions of years ago. All dinosaurs are now extinct.

Erode To wear away.

Extinct Not existing anymore. An animal is extinct when they have all died out.

Fossil Any part of a plant or animal that has been preserved in rock. Also, traces of plants or animals, such as footprints.

Gape A widely opened mouth.

Jurassic The second period of time in the age of the dinosaurs. The Jurassic began about 206 million years ago and ended about 145 million years ago.

Pack A group of hunting animals.

Paleontologist A scientist who studies ancient forms of life, including dinosaurs.

Raptor A type of dinosaur that had a very large claw on each of its back legs.

Reptile A cold-blooded animal, such as a lizard. Dinosaurs were reptiles, too.

Sauropod A type of dinosaur that had a long neck and tail. Sauropods included the largest of all dinosaurs.

Skeleton The bones in an animal's body.

Skull The bones of the head of an animal. The skull does not include the jaw, but many skulls have jaws attached.

Stegosaur A type of dinosaur that had upright plates or spikes growing from its back.

Triassic The first period of time in the age of the dinosaurs. The Triassic began about 248 million years ago and ended about 208 million years ago.

INDEX